Materials

Water

Cassie Mayer

Heinemann
LIBRARY

www.heinemann.co.uk/library
Visit our website to find out more information about Heinemann Library books.

To order:
 Phone 44 (0) 1865 888066
Send a fax to 44 (0) 1865 314091
 Visit the Heinemann Bookshop at www.heinemann.co.uk/library to browse our catalogue and order online.

First published in Great Britain by Heinemann Library,
Halley Court, Jordan Hill, Oxford OX2 8EJ, part of Pearson
Education. Heinemann is a registered trademark of Pearson
Education Ltd.

Editorial: Diyan Leake
Design: Joanna Hinton-Malivoire
Picture research: Tracy Cummins and Heather Mauldin
Production: Duncan Gilbert

Originated by Chroma Graphics (Overseas) Pte Ltd
Printed and bound in China by South China Printing Co. Ltd

ISBN 978 0 431 19263 5
12 11 10 09 08
10 9 8 7 6 5 4 3 2 1

British Library Cataloguing in Publication Data
Mayer, Cassie
Water. - (Materials)
1. Water - Juvenile literature
I. Title
620.1'98

Acknowledgments
The author and publisher are grateful to the following
for permission to reproduce copyright material: © Getty
Images p. **14** (Eiichi Onodera); © Heinemann Raintree
pp. **5**, **7**, **8**, **13**, **16**, **17**, **22** middle, **22** top (David Rigg);
© istockphoto p. **12** (Doug Nelson); © Shutterstock pp. **4**
(Chris Hill), **6** (Karla Caspari), **9** (Shutterstock/pmphoto),
10 (Vera Bogaerts), **11** (Mikhail Olykainen), **15** (Can
Balcioglu), **18** (Angie Chauvin), **19** (Geoffrey Kuchera), **20**
(Sn4ke), **21** (Tony Campbell), **22** bottom (Can Balcioglu),
23 bottom (Mikhail Olykainen),), **23** middle (Chriss Hill),
23 top (Geoffrey Kuchera).

Cover image used with permission of © Jupiter Images
(FoodPix/Susan Kinast). Back cover image used with
permission of © Shutterstock (Vera Bogaerts).

Every effort has been made to contact copyright holders of
any material reproduced in this book. Any omissions will
be rectified in subsequent printings if notice is given to the
publisher.

Contents

What is water? 4

What happens when water
    gets cold? 8

What happens when water
    gets hot? 12

Different forms of water 16

What can water do? 20

Forms of water 22

Picture glossary 23

Index 24

What is water?

Water is a natural material.

It is found in the world around us.

Water is a liquid.

Water is changed by cold.

Water is changed by heat.

What happens when water gets cold?

Water changes when it gets cold.

Water freezes when it is very cold.

Frozen water is called ice.

Ice is a solid.

What happens when water gets hot?

Water changes when it gets hot.

Water boils when it is very hot.

Boiling water turns into steam.

Steam is a gas.

Different forms of water

Water can change form.

Water can be a liquid.

Water can be a solid.

Water can be a gas.

What can water do?

This water froze so it became ice.
It changed from a liquid to a solid.

When it gets warm, the ice melts.
It changes back to a liquid again.

Forms of water

◀ liquid

◀ gas

▲ solid

Picture glossary

 **gas** something that is in the air. A gas has no shape.

 **natural** in the world around us. Plants, animals, rocks, water, and soil are part of the natural world.

 **solid** something that has a shape

Content vocabulary for teachers

material something that can be used to make things

Index

boiling water   14

cold   6

forms   16–19

frozen water   10

gas   15, 19, 22

heat   7

ice   10, 11, 20, 21

liquid   4, 17, 20, 21, 22

nature   4

solid   11, 18, 20, 22

steam   14, 15

Notes for parents and teachers

Before reading   Half-fill a large bowl with lukewarm water. Provide a variety of jugs for pouring. Allow children to experiment with pouring and collecting water. Challenge them to hold water in their hands to demonstrate how difficult it is to hold a liquid. Tell them you can hold water in your hand. Reveal an ice cube in your hand. Explain that ice is water that has frozen. It has become solid, and we can hold solid objects.

After reading

• Almost fill a polythene bag with water and tie it securely. Let children hold the bag and gently squeeze it to see how its shape changes. Freeze the bag of water. Let children wear gloves and feel how it no longer changes shape. Challenge them to think how to make the ice change shape.

• In the hall or outside, fill a bucket of water. Place a coin at the bottom of the bucket. Give children coins and challenge them to use them to cover the coin at the bottom. Encourage the children to observe what happens as the coins drop through the water.

• Place white chrysanthemums in separate glass jars. Add a few drops of different food colouring to each jar. Discuss how the colour in the flowers has changed. Explain that plants draw water up their stem.